BEIJING DAYS

Jon Burris

FOREIGN LANGUAGES PRESS

First Edition 2008

ISBN 978-7-119-05488-9
© Foreign Languages Press, Beijing, China, 2008
Published by Foreign Languages Press
24 Baiwanzhuang Road, Beijing 100037, China
http://www.flp.com.cn
Distributed by China International Book Trading Corporation
35 Chegongzhuang Xilu, Beijing 100044, China
P.O. Box 399, Beijing, China
Printed in the People's Republic of China

Contents

Transition

"One cannot know Beijing in a single visit; not in

two, not in ten. You can only collect impressions,

one upon another, overlapping images in the

mind's eye, fragments of memory that layer upon

layer, comprise a sense of place, of time passing,

and of the future unfolding before your eyes."

Jon Burris

I have always been a photographer, but I was first introduced to Beijing in the summer of 1995 when I went there in my capacity as a curator to interview Chinese artists for a book on contemporary art. For anyone involved in the fine arts, China at that time was one of the most exciting places to be. In particular in 1995, one could sense that a virtual Renaissance was underway and Beijing was at the center of it. In a greater sense, the Renaissance encompassed all levels of Chinese culture - initiating bold new concepts in architecture (angular steel and glass high-rises were becoming the backdrop for tile-roofed temples), in the performing arts and literature (the traditional Peking Opera was sharing the stage with modern dance and performance artists), in the movie industry (fifth generation directors like Zhang Yimou were experimenting with new methods of filmmaking), and not in the least, the Internet was introduced allowing international communication as a result of China's new openness. Perhaps from my own perspective I felt visual artists had experienced the greatest degree of change in the decades following the 'cultural revolution' (1966-1976) and I came to believe that they were working rather feverishly to catch up with the rest of the world whether it be in the pursuit of realist painting or their interpretation of avant-garde art in all mediums. Everywhere I went in Beijing, exciting things were happening; the Central Academy of Fine Arts was enrolling hundreds of students in newly added programs, the National Art Museum of China was undertaking a massive expansion of its exhibition space to accommodate a growing number of visitors, and for the first time in the city's history, commercial art galleries were opening, in part for the benefit of foreigners like myself who were arriving in ever increasing numbers. Of course Beijing's growth was not all occurring in the area of the arts; a booming economy and social restructuring was reshaping the city in virtually *every* way. As extraordinary as the times seemed in the mid-1990s, I could not have predicted the extent to which Beijing would transform itself over the next decade, and frankly, I could not have imagined that I would have been able to witness it to the degree that I have. I consider myself fortunate to have discovered Beijing when I did and for the experiences I have had there over the past thirteen years resulting in the photographs that make up this book.

Regardless of my responsibilities as a curator on my first trip to Beijng, I spent a lot of time exploring the city as a photographer. Undoubtedly, I had come with expectations of what I would see based on memorable images by famous photographers like Henri Cartier-Bresson and Marc Riboud (who I surprisingly ran into one day at the National Museum). I carried with me a mental picture of hundreds of people riding bicycles in the streets or practicing *tai ji quan* (shadow boxing) exercises in the parks. I expected to see kites flying in the great

Tai ji quan exercises in a park, 1995

expanse of Tiananmen Square and I looked forward to exploring the quiet corners of the Forbidden City and the lively back alleys of the hutong neighborhoods. Perhaps it was the China of travel guides that I expected and I was not disappointed with what I found; it was all there. It was the Beijing that was morphing into an ultramodern city that I was surprised to discover. In almost any direction I turned, Beijing was being dismantled and rebuilt, its skyline growing higher and higher. At the time, it seemed like there were a greater number of hotels and banks being built than anything else, but there were also massive apartment complexes, international corporate headquarters, and shopping malls going up in all districts. There was a saying back then that for every construction crane you saw, there would be a new twenty-story building in its place within a month. Looking out of the window of my hotel on my first day in Beijing, I counted over 100 cranes. That was thirteen years ago and today there are over 10,000 new buildings being constructed in Beijing on any given month. For certain, Beijing challenges the sensibilities. It surrounds you not only with its constant flow of humanity but with its contrasts. Ancient temples and watchtowers coexist within blocks of futuristic-looking office complexes. A new large-scale, ultra-modern National Center for the Performing Arts, jokingly called the 'egg' by Beijingers, sits adjacent to and in stark contrast with the 1950s Soviet-style architecture of the Great Hall of the People. Driving east on the broad boulevard of Jianguomenwai Ave past the 'egg' you can blink, and the next thing you see is Tiananmen Square, leaving you with the impression that you have passed from one era into another. On one of my earliest visits to Beijing, I remember a taxi driver proudly pointing out the first McDonald's restaurant in the city. It occupied part of a two-story building just across the corner from the historically important Beijing Hotel and I remember thinking that even though the colors of the golden arches echoed the imperial reds and yellows of the Forbidden City, it was somehow disturbing in this landscape and I could have done without such homogenization. During the same trip I stayed in a new joint venture hotel a few blocks to the north. As modern as it was and with great views—to the west you could make out the gates of the Forbidden City and to the south, McDonald's – I most enjoyed looking

National Center for the Performing Arts (horizontal image), 2007

out the window of my 24th floor room and down on an old wood and tile-roofed neighborhood that surrounded the Central Academy of Fine Arts, which was still active in that location. I could see outdoor food vendors cooking in woks and bicycle carts delivering fresh vegetables. I could also see a multi-level mall that was being built on the closest corner and I was informed by my hotel that probably on my next stay with them, the Central Academy would be gone and I would have convenient access to modern restaurants and shopping centers. For something gained, something is lost; such is progress. The Central Academy moved to a new home in the far north of Beijing and more than quadrupled its space, adding a beautiful museum and far better student housing. The first McDonald's was leveled to make room for another shopping mall,

Rooftops of a hutong neighborhood, 2005

the Oriental Plaza. On a recent trip to Beijing, I was in a taxi headed east on Xichang'an Jie and as I passed the corner where McDonald's had been, I looked north beyond the Oriental Plaza and saw what appeared to be an exact replica of the George V Hotel in Paris. I blinked, but I was still in Beijing.

Part of what one learns about Beijing is that pretty much since the late 1980s, enabled by Chinese leader Deng Xiaoping's sweeping economic reforms, the city has remained under constant renovation and construction. Everyone from artists, to businessmen, to civic leaders have entered into a search for new ways to interpret the Beijing of the 21st century. I cannot think of a model for such unprecedented growth and creative energy anywhere else in the world. What one often hears today

A tour group poses for pictures in front of the Beijing Olympic stadium, better known as the Bird's Nest, 2008

is that the estimated 40 billion dollars being spent on updating Beijing can be attributed as well to the city's winning the bid to host the 2008 Olympics which will also be something of a coming-out party for the new China. Maintaining its lead in innovative architecture, the focal point of Beijing's central Olympic village has become an iconographic stadium

Dongbian Men Gate, 2008

design by the Swiss architectural firm Herzog & de Meuron. Because the Chinese have a tendency to attribute new forms of architecture to familiar subjects, the stadium is commonly referred to as the 'Bird's Nest' due to its abstract similarity to intertwined twigs. In the early stages of conceptualizing the stadium, the well-known Beijing avant-garde artist Ai Weiwei was a consultant to the architects and, in view of his practice of sculpturally abstracting and incorporating Chinese objects like Ming furniture and ceramics into his art installations for international audiences to "reconsider", it makes sense that he would propose the

Bird's Nest as an influence and that it would become a popular symbol. Ever since it began to take recognizable shape, Chinese artists working in all mediums have referenced it and commercial businesses have turned it into something of a logo, so much so that today the image of the Bird's Nest is hard to ignore given any consideration of the new Beijing.

In addition to the Olympic undertaking, renovation projects are ongoing at all of the city's historical sites with monuments covered in acres of the familiar green sheeting that is used to keep dust and debris in check (I once thought, how is this different from Europe as the last time I was in Paris, Notre-Dame was enveloped in scaffolding). However, some of the most impressive new additions to Beijing can be found in the CBD (Central Business District). It is nothing short of an experimental playground for a host of international architects all of whom are able to see their most elaborate designs come to life given lower construction costs in China (such building projects in New York would cost fifteen times as much to complete). At the center of the CBD are two major skyscrapers; the first is the impressive future home for CCTV (China Central Television). A 6.5-million-square-foot structure consisting of two 768-foot offset towers joined by a cross-section, the CCTV building was designed by the Dutch architect Rem Koolhaas. The second stand-out building in the same neighborhood is the 74-story China World Trade Center Tower created by the American architectural firm Skidmore,

Owings & Merrill. As a photographer searching for a way to show the vast area these two buildings occupy, I relied on the aid of my friend and sometimes fellow expeditionary, Han Weiqiang, his financial consultancy firm offices being in Tower One of the China World Trade Center. He was able to get me access to the roof of his building from which I made the only panoramic photograph that appears in this book. This view of Beijing looking northeast gives one the feeling that it is as cosmopolitan as any other international city but at the same time, maybe a little futuristic. No doubt in another decade the skyline will have changed just as dramatically.

I would not want to leave the impression that Beijing is only concerned with modernization because that is clearly not what I have found. There is a disquieting tendency among a number of China watchers to decree the "demolition" of the historical capital and to contend that the remodeling of the city has rendered obsolete important parts of it. The fact is Beijing is evolving so quickly that it is easy to get lost on a street you have traveled down dozens of times. You look for the small corner tea shop you remembered as a landmark but it's been moved, replaced by a Starbucks. The eight-level gray concrete apartment buildings that always had laundry waving from their windows have been rebuilt into thirty-story modern-looking residential/office 'communities' with Western-influenced names like SOHO, Fifth Ave., or MOMA.

However, just as you begin to wonder where you are, you catch sight of a hand-carved wooden eave and ceramic tiled roof and you know you're still in Beijing. You turn down a street lined with willow trees, the rush of yellow/green taxis and Passats with dark tinted windows disappears and you see bicycles peddled by parents shuttling children to school as they have done decade upon decade. A street vendor prepares dumplings in a wok for people on their way to work and a coal brick seller pushes his cart down the sidewalk selling fuel to clerks door to door. This is Beijing as it has always been. You may be aware that the future is developing right in front of you but reminders of the past never completely disappear and this is what helps define the character of the city.

As a photographer it has quite literally taken me over a decade to arrive at a personal interpretation of all that Beijing is and obviously my photographs fall within the context of my own experiences. While the very nature of photography dictates that all photographs are documents, I hasten to point out that I am not a 'documentary' photographer in the sense that Henri Cartier-Bresson and Marc Riboud were. Although the viewer will find a lot of photographs of people within this book, I do not consider myself to be a social observer so much as an environmental observer. Just as I've watched Beijing evolve, I've watched my own photographs evolve into something more than singular images and I've come to embrace the idea that seemingly unrelated images may communicate more clearly when combined. In joining two photographs

A panoramic view of new construction from the roof of the China World Trade Center Tower, 2008

the meaning of each may be changed so, in effect, a third image is created making for a completely different statement. It is in this act of joining images that I am separated from the documentary tradition that seeks

Transition photograph, 2008

primarily to preserve that 'decisive moment' before the camera; to freeze an image of a place in time. Instead, I have chosen a more interpretive approach with which to view Beijing. It is a way of presenting images that challenge the viewer to consider more than the moment; to view a city in transition, to be able to place images layer upon layer until you can perceive the entirety of what is in front of you. It is but a different way of organizing the past and present for the here and now. When I look at photographs I made during my first trip to Beijing in 1995, I realize that I was still working predominantly in black and white and with film instead of digital files, not that technology necessarily has an influence on the way I view the world. My focus then was more on street life and what I would call the intimate landscape. I was far from contemplating the collaging of images or how to represent Beijing's spiraling growth; I seemed to be looking down, not up! One of my favorite images from this time is of the Dongbianmen Watchtower that I made with a small plastic

Dongbian Men Gate Watchtower, home of the Red Gate gallery, 1995

'Bird's Nest' stadium, 2008

toy camera called a Holga. With no control over focus or exposure, it produces a softened photograph that I find quite appealing in its simplicity and I think the process fits the subject, the image of which has appeared in numerous historical examples. Recently, while trying to arrive at an approach to photographing the Bird's Nest stadium – something

Morning market, 1995

of a parting image for the book – I thought again of the Holga. In my photograph of the stadium, I abstracted the structure because practically all photographs I have seen of it include it in its entirety. I like the resulting image a lot because I feel that somehow across time, there is a relationship between the Dongbianmen Watchtower and the Bird's Nest stadium. They will likely be remembered as symbols of Beijing; at least both landmarks made a lasting impression on me!

19

A decade ago my response to making photographs of Beijing was decidedly different; today I see it through the filter of time passed. In my earlier Beijing days, I enjoyed getting up before dawn and going to the street market set up close to the Meridian Gate of the Forbidden City.

Rooftops of the Forbidden City, 2008

Everyone came there to buy the day's best vegetables or fish. You could watch small groups exercise or couples learning ballroom dancing. For pocket change you could get a haircut or a massage. Today, however, the Forbidden City market, like most small street markets, has disappeared and groups who exercise or dance have moved to large public parks. There are more grocery stores now than street markets, more hairdressing salons than independent barbers. Inside the Forbidden City (one of my favorite places to photograph) on the occasional quiet day when there are not as many tourists, I can still escape the clamor of the city; I can feel a palpable sense of history there. I can find the abandoned apartment where an artist I know lived without electricity or running water during the 'cultural revolution' and it still has no electricity or running water. In anticipation of the Olympics, the main palaces and halls of Beijing's most visited tourist attraction have been renovated; the dust has been cleared from the throne rooms and the ancient, broken stones in the palace stairways have been repaired.

The past has not disappeared though; it is still there for those who take the time to relate to it. The Forbidden City has existed in Beijing (previously Peking, previously Yanjing) since the 15th century and it will hopefully exist for many more centuries, perhaps even after the Olympic village has disappeared.

The book on artists that brought me to Beijing in 1995 is long since finished but I have remained in contact with the Chinese art world because I am still a curator. In the 1990's, I used to visit artists in their government-assigned apartments where they carried on their careers in the same space that they raised their families. At that time, many were still teachers working in government-placed jobs in the fine art academies. Some functioned as set painters on army movie lots or mural designers. All of them produced personal art in their spare time with the

Zhang Xiaogang in his studio, 2008

goal of becoming what they called 'professional' artists, living totally on

the sale of their work to collectors and museums. In a few of my portraits

one cannot help but notice how their 'studio' is simply a small corner

filled with paints and canvases. I always wondered how they accomplished

what they did under those conditions. Change (and attention) came

quickly though in the heady days of the post-'cultural art revolution'

of the 1990s and today, many of the same artists I photographed have

become successful far beyond the dreams of even artists in the West. The

contemporary Chinese artist's works command almost unfathomable

prices at auctions worldwide and there are waiting lists of collectors who want their paintings at any cost. Most of the artists whom I met a decade ago have since moved to one of the more than thirty surrounding communities that have developed just outside of Beijing; in the West we would call these suburbs. They have built large homes with enormous studios where they now have plenty of room to paint, and store their canvases, and entertain a steady stream of interested parties like myself. The 'artist villages', as they are referred to, are also a part of the burgeoning capital city and have become locally symbols of a new class of successful and wealthy Beijingers.

Over time my mental image of Beijing has developed through an engagement with the past and present. In fact, I have come to believe the only way one can truly gain a sense of identity for this city is by incorporating the past with the present, as you will see in the photographs that comprise the following chapter entitled *Transition*. In the process of conceiving them, I reflected on something the critic John Berger wrote, "All photographs are there to remind us of what we forget." After thirteen years of experiencing Beijing I know that it is not only what one sees in guidebooks (although I don't want to forget that part of it either), it has so many more layers and so much to offer. I believe that it is possible to accept what comes with change and progress without giving up an appreciation of the past. If *every* hutong neighborhood were leveled to make way for yet another high-rise, I would join the protests

Transition photograph, 2008

against such progress, but that's simply not the case. Change is necessary and why would anyone expect less than *dramatic* change from one of the most dynamic cultures and one of the most vibrant cities in history? I will go as far as to say that to understand China, you must begin with Beijing. I have personally discovered so much there and I anticipate far more in the future. When I am in Beijing, I sometimes miss my home, but when I am home again, I *always* miss Beijing.

Living in Beijing

"Although recalling the past may make you happy,

it may sometimes also make you lonely, and

there is no point in clinging in spirit to lonely

bygone days. However, my trouble is that I cannot

forget completely, and these stories have resulted

from what I have been unable to erase from my

memory."

Lu Xun - Call to Arms

I came to Beijing in 1995 to learn about its artists, about their lives in transition and all that influenced them during this incredibly important period of change in China. Given the atmosphere of opening and reform (begun in late 1978) many artists I met were choosing to interpret freely the normal, everyday aspects of their lives, so I went in search of this myself, finding in the process that what contributes so greatly to the character of Beijing is its people. The faces I found in Beijing became as important to me as the details of its architecture, old and new.

Throughout the capital's long history, much of everyday life has taken place in and around its most central landmark, the Forbidden City. In 1995, I was told that one of the best places to experience the "face of Beijing" was at the early morning market just outside the Meridian Gate entrance of the Forbidden City that opened there at sunrise and lasted only for a couple of hours. I went to it several times, fascinated by everything that took place surrounding the market. Numerous barbers set up shop on the sidewalks with little more than a chair and a pair of scissors and they always had customers. You could also get a quick massage or join in group *tai ji quan* exercises. Bicycle cart merchants sold vegetables, fruit, noodles, spices, bread or tea and a variety of fresh fish flopped around in tin pans. People wandered about, casually shopping and visiting and in those early days, I was able to move through the market leisurely photographing what would become a dated life-style. I was surprised at how easy it was to get to know Beijingers by visiting places like this. Within minutes of starting any conversation, you could know how long someone had lived in the neighborhood and their children and

A view down a corridor in the Forbidden City, 1995

grandchildren's names. Almost always, a paper cup of tea appeared out of nowhere and I found myself sitting and talking as best I could through the language barrier. While making a photograph of a young boy getting a haircut, I learned from his grandfather (watching in the background) that he took care of the child every day while the boy's parents – the man's son and daughter-in-law – both went off to sales jobs at one of Beijing's new shopping malls. He explained how very different this was from the way he had grown up and how different his grandson's life was surely going to be when places like the markets disappeared.

Two photographs made during early visits *inside* the Forbidden City are of special interest. In one, a grainy black and white image, I was looking through an open door and down a corridor into a scene that seemed out of ancient history – just as I had expected the interior of the Forbidden City to look – when suddenly

A long time resident of a Forbidden City apartment, 1995

four Chinese men dressed in modern suits came strolling out of another doorway into the picture; I assumed they were tourists. I am still struck by how anachronistic an image this is. Another photograph made on the same day is of a woman sitting in front of a caretakers' apartment. She is wearing Mao-era clothing, typical of an older generation of Chinese.

Behind her hang two long strands of garlic and a string of firecrackers; at her feet is a stack of sweet potatoes. I remember talking to the woman who explained that she had been a caretaker and gardener inside the Forbidden City most of her life. There are no longer such jobs at the Forbidden City and this photograph simply could not be made there today. It truly is a part of history.

All of my early portraits of Beijingers hold different meanings for me personally; an old man sitting in a chair in the street in front of his hutong home is as representative of Beijingers as the pedicab driver looking out the back window of his customized taxi. Two gentlemen comparing their pet birds are as typical as a trio of retirees doing synchronized exercises in the park. If you look long enough, you can still find Beijingers doing the same things, however, they are joined now by a newer generation who talk on cell phones while riding to work on motor bikes, or rush in and out of designer stores like Burberry and Ermenegildo Zegna.

Before I went to Beijing, I was inspired by Bernardo Bertolucci's beautiful film *The Last Emperor*, the first, and one of very few films actually made on location inside the Forbidden City. I found, however, that trying to interpret the 'landscape' of the Forbidden City was difficult. The expanse of the palace with its courtyards widely separating its halls and pavilions was so vast. I decided that to do this properly meant choosing the exact time of day -- for example, when the late afternoon light stretched shadows and gave the buildings more definition. The problem was, that usually ended up being about an hour before closing time to tourists and I felt rushed. I made attempts at photographing there during my first two or three trips but I was never quite satisfied with the results. My viewpoint changed, however, when in the course of

A busy couple on a scooter, 2008

my curatorial work, I arranged to meet the artist Li Kai for an interview about his paintings, all of which focused on the Forbidden City. He had been assigned to live there during the 'cultural revolution' to work as an art restorer and he occupied a small apartment in an area the general public never sees. He took me on a tour of his private Forbidden City to show me where he had made specific paintings, especially one of an old weathered door that he passed by daily. That one incident changed my perspective. After that, I became interested in details. I photographed rather obsessively a wide variety of stone gargoyles, dragons and lions, bronze animals of all types, and hand-carved wooden roof guardians (created to protect buildings from fire). Rooftops became a special interest in fact. From certain angles, I could relate to why their curving style and upturned corners had been modeled on the wings of birds and I began to look differently upon modern architecture that referenced these ancient rooflines (somewhere around the mid-1990s foreign architects drew upon such details for new building designs, usually hotels).

Interestingly, one day, quite by accident, I stumbled into a small room in the Forbidden City and found a modest display of 'the last emperor' Puyi's bicycle and a pair of sunglasses, both given to him by Reginald F. Johnston, the British colonial official who became Puyi's tutor in 1919 following the young emperor's abdication from the throne. As iconographic elements in Bertolucci's film, these relics from history were as fascinating to me as any artifacts in any of the palace museums!

In a way, I think of all of my photographs made in Beijing as observations recorded in an ongoing abstract journal. To me a stone lion inside the Forbidden City has a relationship to a stone lion that stands in front of Beijing's largest bank in the financial district. Across

Sculptural details, 1995

Viewers at a fine art auction preview, 2008

time, they both represent power – power that existed 500 years ago and power that exists today. I find the juxtapositioning of ancient and modern architecture in Beijing as interesting as any place in the world. There is a certain harmony in the way one building is situated against another, quite possibly because of the practice of *feng shui* alignment which is still respected even by modern international architects. A good example of this is in my photograph of Nanxincang, the imperial granary whose ancient tile roofs and post and beam entrances have been preserved and now sit, surrounded quite closely and in stark contrast to, colorful high-rise office buildings of steel and glass. Like many such places throughout the city, this historical location has been 'repurposed' and now comprises of small art galleries and restaurants. One day I sat in one of the restaurants and looked out on the new Beijing skyline and I remembered feeling quite at peace with the past and present represented together. As Beijing enters the 21st century it is becoming a city of extreme contrast, reflected not only by its architecture, but by what it has to offer. You can bargain for the purchase of antique furniture, Mao memorabilia, gramaphones, shadow puppets, and ceramics in outdoor markets like the one at Panjiayuan that are as colorful and diverse as any to be found throughout the world. Or, you can preview both historical and contemporary art at any one of Beijing's dozen or more auctions held twice yearly that regularly sell works in the millions of dollars. At one time, travelers used to come to Beijing for 'Peking duck' but now they find international offerings, a fusion of Chinese, Italian, French or other cuisines in eloquent settings like the Philippe Starck designed *Lan Club*, or the Hyatt's exclusive *Made in China*.

Ten years ago, you could look out across any one of Beijing's expansive boulevards and see large masses of bicycles moving rhythmically like a school of fish in the ocean. Today there are more cars than bicycles and the ocean looks more like a rushing stream with single fish darting one way or another to avoid colliding with the cars. The thing is, bicycles haven't disappeared from the Beijing landscape; they

A dance instructor with a couple of students in a public park, 2008

are still a part of life in the city. Fewer hutong neighborhoods exist and there are far more skyscrapers today than there were ten years ago, but you can still find the hutongs. While the Forbidden City morning market has ceased to be, one can still find street markets throughout Beijing. People still gather in public parks to play mahjong or take dance lessons; they still carry on business in small corner stores and buy breakfast or lunch from bicycle carts. These are just as much the rituals of daily life as they've ever been and I've tried to gather impressions of it all because this is life in Beijing. This historical city, this modern city, will no doubt continue to remake itself far into the 21st century. If Beijing's tallest building, the 768-foot CCTV tower, seems futuristic today, wait until 2050! I can only imagine looking down from the perspective of a 100-story office tower then, at bicycles moving in unison past Tiananmen Square just as I remembered them in 1995 and 2008.

Early morning exercises in front of a wall of the Forbidden City, 1995

Two retired gentlemen with their
pet birds, 1995

[Left] A quiet corner of the Gate of Supreme Harmony, Forbidden City, 1995

[Right] Rooftop guardians, Forbidden City, 2007

[Left] A wall in the Summer Palace, 2004

[Right] A lantern in the Forbidden City, 2004

[Left] Gargoyle, Forbidden City, 2007

[Right] Sunset reflected in moat at the Forbidden City, 2007

Kunming Lake, Summer Palace, 1995

[Left] Entryway to two apartments in a Beijing hutong neighborhood, 2004

[Right] Architectural detail of a building in the Hou Hai district, 2008

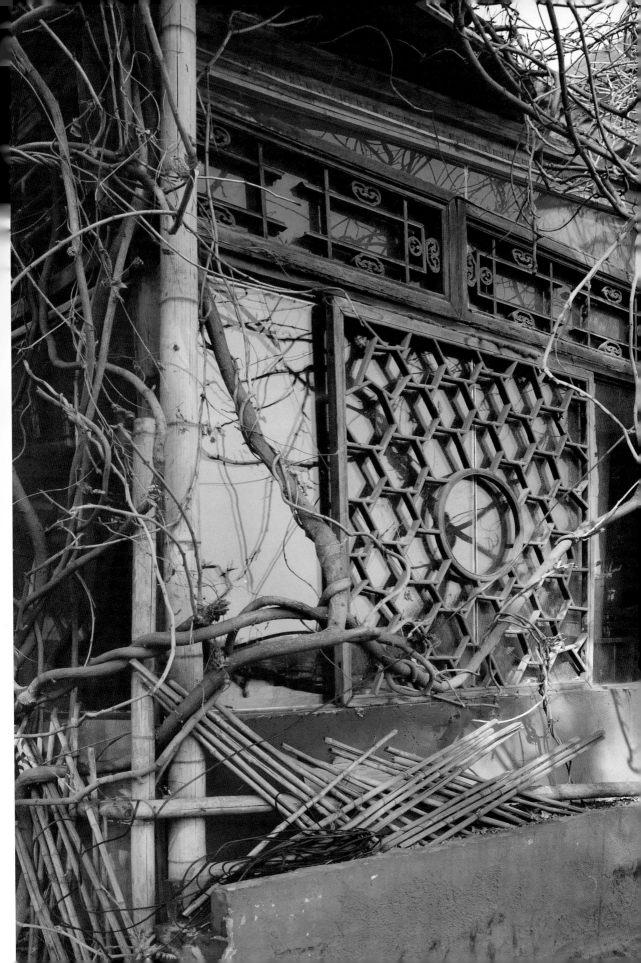

[Left] A barber gives a haircut on a sidewalk near the Forbidden City, 1995

[Right] Morning market, 1995

[Left] A young cook at an outdoor restaurant, 2004

[Right] A retired man sits in the street in front of his hutong home, 1995

Bei Hai Park, 2005

Two women take a pedicab tour through a hutong neighborhood, 2008

Wooden statues at the Panjiayuan antiques market, 2008

Acupuncture models in an antiques market, 2007

[Left] Contrasting view of old and new architecture, 2008

[Right] Skyscrapers and construction cranes, 2008

[Left] Stone lion in the Forbidden City, 2007

[Right] Financial District, 2008

[Left] Two young shoppers entering a store at Oriental Plaza, 2008

[Right] A young soldier poses for a photograph in front of a bronze statue in the Forbidden City, 1995

A view of the Beijing skyline from inside the New Poly Plaza, 2008

Entry to Tsinghua Science Park, 2008

Bicycles pause at a busy intersection, 2008

A Lamborghini dealership, 2008

浪琴表《名匠》系列

浪琴表北京办事处: 010-6523

Traffic in a late afternoon rain storm along Jianguo Men Nei Dajie, 2004

Modern shopping complex, 2008

[Left] New China World office tower under construction, 2008

[Right] Signage for a gallery in an office building, 2008

[Left] A young woman serves tea in a tea house, 2004

[Right] Yang Jie (my associate and translator), 2004

[Left] Six young men on a lunch break, 2004

[Right] A gentleman on a bicycle, 2008

[Left] A woman sits with her dog in front of her small grocery store, 2008

[Right] Four young actresses take a break during filmmaking in the Forbidden City, 2004

[Left] A pedicab driver with his customized cab, 1995

[Right] Two student athletes on bicycles, 2008

A photographer instructs a new bride on how to pose along a canal in the Hou Hai district, 2008

A common scene of a young mother transporting her child on a bicycle, 2008

[Left] A bookseller at an antiques market, 2005

[Right] An early morning dominoes game, 2008

A tour group prepares to enter the Beijing Olympic stadium, better known as the Bird's Nest, 2008

Today's Art

"In 2007, China displaced France as the art
world's third-biggest auction market after the
United States and Britain. There is no other
market that compares to the Chinese art market
in growth."

Jon Burris

Just as Beijing has experienced some of the most profound changes in its history, especially over the last two decades, so has the Chinese art world. As far as I am concerned, there is nothing to compare it to in 20th century art history. The rise of Chinese art, and the success contemporary Chinese artists have experienced has been nothing short of phenomenal. While it may be arguable to a few well-known artists in Shanghai and from Sichuan province, most of those involved in the fine arts believe that Beijing remains the heart of the Chinese art world and the place of greatest opportunity for artists.

Off and on over a period of three months in 1995, I spent my early Beijing days meeting artists in their studios for interviews and photographs. Typically their studios were just a corner of a small cramped apartment that they shared with their wives, their children, and sometimes their parents as can be seen in my photograph of artist Wang Yidong, painting. It was often difficult for me to make one 'studio' look different from another as all apartments pretty much looked alike back then. However, like all things related to the Chinese art world, that would change dramatically over the next thirteen years and I would eventually find myself stepping into vast, high-ceilinged spaces reserved only for work and separated from living quarters that were often ten times the size of the artists first apartments. A good example of this is Zhang Xiaogang's new studio I visited recently that looks more like the inside of an airplane hangar! It is wonderfully spacious and perfectly laid out for access to his predominantly large canvases. When I photographed him in it, we laughed at how different his life was from the first time we met in 2001. He guessed that the old apartment was 600 square feet and he joked that he could only work on one or two paintings at a time there.

A lot of the artists I had occasion to meet in the early days were from the first post-'cultural revolution' graduating classes of the 1980s. Some held teaching positions at the art academies, others worked in government-assigned jobs, although all of them expressed a desire to eventually become 'professional' artists pursuing only private work. It was hard then to know which ones would succeed, amazingly, all of them did! I was fortunate, too, to meet a few of the most respected Chinese artists from the first generation of oil painters -- those who had been able to travel outside of the country to places like Japan and France to study with the idea of returning home to bring new ideas into the academies.

I've always liked my photograph of Wu Zuoren, one of the most respected of the first generation of Chinese oil painters. It turned out

Wu Zuoren

to be the last serious portrait made of him as he died shortly after our

meeting in 1997. There is nothing in the photograph to imply that he

is an artist. He had long stopped painting by the time we met for an

interview in a park outside of his home in far northwest Beijing. There

is something in his gaze, however, that left me with the feeling he was

studying what I was doing, as if at any moment, he would instruct

me on the best angle to choose or the way to use the mottled light

coming through the trees. My portrait of another highly respected first generation oil painter, Wu Guanzhong remains a favorite as well. He was 77 in 1996 when I photographed him in the small bedroom studio of his Beijing apartment that he still works in today. At that time, Mr. Wu was beginning to receive international recognition from institutional and private collectors and numerous publications on his work began to appear. As an immensely influential teacher, he was unquestionably the first credited with encouraging Chinese artists to develop their own styles, even if they utilized Western techniques. When we went into his

Chen Yifei

studio, he sat a piece of watercolor paper in front of him with the idea of doing a demonstration; however, we became so involved in a conversation covering everything from his difficulties during the 'cultural revolution', to his surprise at how quickly the market for Chinese art was developing, that he never got around to painting and I have always liked the blank canvas, as it were, in front of him. It is appreciable only if you know how prolific an artist Wu Guanzhong is! Some of the Chinese artists spanned other careers as designers or architects, or even filmmakers. One was Chen Yifei who was actually from Shanghai but maintained an office in Beijing. Facilitated by my most determined and capable associate Yang Jie, with whom I have worked throughout the course of all of my thirteen years in China, an interview was set up in my room at the Palace Hotel. Yifei, who by 1995 had become a recognized film director, was passing through Beijing on his way to the Cannes Film Festival and happened to be staying in my hotel. I will never forget my introduction to him in the lobby. Traveling with him was the beautiful actress, Gong Li, who appeared in the movie he was about to preview in Cannes. Once she was spotted by fans we had to escape to my room and I was left to try and bring some order to the interview and photo session with Yifei. As it turned out, it was a very good interview and the start of a long acquaintance with the artist. I believe in my portrait of him, Yifei looks sophisticated -- every bit the film director, but also extremely at ease, more so than I think I was at the time.

In the mid-1990s when Chinese artists in Beijing did sell paintings directly to private collectors or museums – or more rarely through galleries – it was usually for a few hundred dollars. There were very few public galleries in Beijing at that time (most notably the Red Gate

Liquor Factory arts district entrance, 2007

Gallery that is still located in the Dongbianmen Watchtower) and only one or two auction companies that were organizing contemporary art sales. In short, it was still very much a developing market. In two or three years though, all of that started to change. Some artists like Yue Minjun had begun to move to small villages like Yuanmingyuan just outside of Beijing where they occupied abandoned farmer's homes. They rarely lived in these places fulltime, but they had plenty of room for canvases that seemed to be growing in size proportionate to their careers. As they became increasingly prosperous, several of the artists bought multiple homes or apartments and moved casually between them, painting one day in one studio and the next day in another.

One of the more positive developments in the Beijing arts community began in early 2000 when designated 'arts districts' started to appear. The first, and still perhaps the most important, was 798 in an abandoned government manufacturing plant. It became the center for a series of galleries, art studios and coffee shops that sprang up in a gated neighborhood of factories in the Dashanzi

A street in the 798 arts district, 2007

district located just outside of Beijing's Fourth Ring Road. Today, there are over 100 galleries covering two million square feet in Dashanzi alone and there are several competitive districts not far away. Ten minutes to the north of 798 is the *Caochangdi* district, home to the well-run *China Art Archives* and twenty minutes from there is *The Liquor Factory*, so named because it was once the location of a brewery. None of these arts districts existed five years previously. I believe one of the most interesting developments of late is the *Songzhuang Artist's* Village. In a small town an hour east of Beijing, several artists have built homes and studios and the entire community has embraced their arrival, opening art supply stores, canvas stretching shops, galleries, coffeehouses and, within the past year, a very well appointed arts center. While such places exist outside of central Beijing, they are still a very important part of the city's arts community that continues to attract the attention of visitors and collectors world-wide.

Wu Guanzhong

Wu Guanzhong is one of the most respected painters in all of China and certainly one of its most prolific artists and highly regarded teachers. In his art, he has skillfully infused Western oil painting techniques with the delicacy of Asian brushwork and the nuance of traditional Asian water color to create a blended style that is at once modern and unique.

Mr. Wu studied under Lin Fengmian and Wu Dayu, both well-known members of the first generation of Chinese oil painters. In turn, he has had enormous influence on a following generation of artists. It is impossible to imagine contemporary Chinese art without Wu Guanzhong's contributions. While he has traveled extensively and been accorded countless exhibitions internationally, he has maintained a small and intimate studio in his Beijing apartment for decades. This is where my portrait of him was made in the fall of 1996. I think it is evident that he is at ease in this atmosphere and frankly, it would be hard to conceive of him anywhere else.

During the 'cultural revolution', Li Kai was assigned to a job as an art restorer at the Forbidden City. He lived in a small apartment that had no running water and no electricity. Taking inspiration from his surroundings, he began to make paintings of the architecture of the palaces, and in particular he focused on details that anyone other than an artist might overlook. Following the 'cultural revolution', Li Kai was able to establish a career painting this subject only and he is today recognized for his unique and intimate studies of one of Beijing's most well-known landmarks.

When I met Li Kai he offered to take me to the Forbidden City to show me the location of some of his paintings. In particular, I asked to see the ancient apartment where he had been assigned to live during the 'cultural revolution'. It was hard to believe anyone could have occupied the space he showed me, yet as I listened to Kai talk about his days there, I sensed that he had strong feelings of nostalgia about it. I think it is evident in this portrait of him sitting on a flat-bed tricycle in front of the entrance to his former residence.

Wang Yidong is today one of China's most successful oil painters. In the late 1980's he gained attention for a series of paintings of young Chinese women that were compared to the works of Italian Renaissance artists.

I photographed Yidong in his small Beijing apartment in 1995 where his 'studio' was nothing more than a corner of his living room. He seemed happy and told me it was ok working there, but he hoped to one day have more space. Roughly two years later when I saw him again, he had moved to a large English Tudor-styled home in a suburb of Beijing.

Within it he had built a studio the size of a gymnasium. Space was no longer an issue.

Yang Feiyun

Yang Feiyun is known among collectors as a highly-sought-after member of a select group of China's Realist painters who have risen to fame over the past two decades. He is also a respected teacher.

I had photographed Yang previously in a very small studio he used at the Central Academy where he taught in the 1990's. It should be noted that none of the Chinese artists I was introduced to early in their careers complained about their lack of space, which by comparison to artists in the West was absolutely claustrophobic. They simply made do with what they had available.

Not long after we first met, Yang Feiyun moved to a spacious and comfortable studio in a home he built outside of Beijing in the late 1990's. When I visited him there it felt very European to me with aged wooden floors, antique furniture and soft, light filtering through louvered skylights. For a painter who creates classical Realist figure studies, it is a perfect setting.

Ai Xuan is the son of one of China's most famous poets, Ai Qing, and he is the brother of one of its most well known avant-garde artists, Ai Weiwei.

During the 'cultural revolution', Xuan was sent to Tibet and it was there that he decided the subject for his own art would become the faces of the Tibetan people. The style of painting he created for himself was influenced in part by the American artist Andrew Wyeth whom he met in the U.S. in the late 1980s.

Today, Ai Xuan is one of the most recognized and collected of the Chinese Realist painters. I have photographed Ai Xuan in numerous studios over the past thirteen years. In most of my portraits of him, he is characteristically joking or laughing for the camera, but on this particular day I asked Xuan if he would think of what it was like to be one of the subjects of his paintings in the isolated landscape of Tibet. He immediately lost himself in thought and I was able to get the atypical portrait of him that I wanted. It was one of those moments you hope for as a photographer.

Jiang Guofang is a very serious artist whose paintings deal primarily with the history of the Forbidden City and the succession of emperors who lived there. His work has been exhibited in museums internationally and is a part of important private collections world-wide.

Hearing about the thoroughness of his research into his subjects, I was surprised on my first visit to his studio in 2003 to find such a jovial man. I couldn't resist posing him in front of what would become one of his most famous paintings, simply to contrast his attire and demeanor with that of the subjects in the royal court scene he had depicted.

Artist Wang Huaiqing has long been one of my favorite subjects to photograph, in part because I have appreciated his unique style of abstract painting that draws upon the subjects of Chinese furniture and architecture, but also because he is one of the most intelligent artists I have ever met.

In numerous visits to his studios, I have always found the most appropriate backdrops for my portraits to be his large canvases. Because many of his paintings are rendered in shades of black, white, and gray, they work perfectly for black and white photographs. However, in a rare exception, I chose to make this image in color because I liked the way Huaiqing's clothing was set in subtle contrast to his painting.

Xin Dongwang is something of a rising star among the Beijing Realist painters although his work has been widely exhibited over the past ten years. Today, his straightforward interpretations of common people like the Olympic stadium laborers in the painting in this photograph are gaining the attention of critics and collectors alike. He is also a full-time teacher.

Ten years ago if I visited artists who were also teachers and they used the studios the fine art academies provided them, I usually found small, dimly lit spaces. Recently, when I met Xin Dongwang at his studio in Tsinghua University, I was surprised to find a large, two-story open space with a beautiful skylight. His paintings were stacked against the walls and his most recent work was on a rolling easel that he could move around the room to take advantage of the light as it changed throughout the day. The conditions were ideal for the portrait I wanted: the artist surrounded by his work.

Cao Li is an artist whose work has long been of interest to collectors of contemporary Chinese art, primarily because of its symbolism, but also because of its whimsical and stylized nature.

I have been fortunate to have maintained a close relationship with Li since we first met in 1995. I like this portrait of him because I know he is laughing at my reminding him that when I began photographing him, thirteen years ago, he had hair!

Shi Guorui is quickly becoming one of the most recognized Chinese fine art photographers. His unique photographs are made without a camera or film by turning a hotel room or rental truck (even a watchtower on the Great Wall) into a 'camera obscura.' He blackens out the enclosure with sheets of dark plastic and positions photo paper on one side of the enclosure while creating a small pin-hole on the opposite side through which an image from outside is projected onto the paper, exposing it for sometimes several hours thus creating a photographic image that is reversed in tone and backwards, just as it would appear on a normal piece of film.

I particularly like the way the steel rafters of the ceiling repeat the framework of the Bird's Nest stadium in the photograph and the artist himself appears casual and relaxed which fits his personality. Posing artists can often be difficult but I always try to make them feel at ease which fortunately for most of them is typical.

Zhang Fangbai has established himself as an Expressionist painter whose subjects include everything from birds of prey, to pagodas, to iconographic Chinese figures like Mao or Deng Xiaoping. He has held exhibitions in Europe and the U.S. and his work often appears in the Chinese contemporary art auctions.

As a photographer I am excited when I come across unexpected subjects. In 2004, I visited a newly developed artist's community in far northeast Beijing and discovered Zhang Fangbai's studio. I wandered in while he was practicing calligraphy with traditional brush and ink techniques. The contrast of his work-table with one of his Expressionist paintings of a bird hanging in the background, both separated by a modern, bright red sofa, formed an interesting composition. It is also perfectly representative of today's Chinese artists who maintain a respect for their cultural background while embracing modern influences.

Corkboard in Zhang Xiaogang's studio

Zhang Xiaogang is currently the third highest-selling living artist. His 'Bloodline' series of paintings inspired by Chinese family portraits have gained him international acclaim and put him at the top of the contemporary Chinese art world. More recently, he has begun to utilize singular figures in austere interior settings.

In 2000, I visited Xiaogang in a small apartment in Beijing that functioned as his studio. It was empty as he had just sent several paintings to an exhibition in South America. Nothing remained on the walls except a small bulletin board on which was taped several snapshots; there were close-ups of young children, various family pictures and a couple of drawings. I made a picture of that. In April of 2008 I went to see Xiaogang in his new studio on the outskirts of Beijing. This one resembled an airplane hangar in size and there were numerous paintings stacked around the walls. As I walked around exploring the studio, I came to a large, free-standing corkboard with dozens of photographs and drawings taped to it. In the center were the same snapshots I had seen eight years earlier. I pointed this out to Xiaogang and he said, "See, nothing much has changed!" The irony is of course, that much has changed for Zhang Xiaogang and all of the Chinese artists.

Yue Minjun is one of China's top selling artists whose work is universally known, and collected at prices in the millions of dollars. His bald, laughing man (a self-portrait) has become an iconographic symbol of contemporary Chinese art just as Andy Warhol's portraits of Marilyn Monroe became strongly identified with American Pop art.

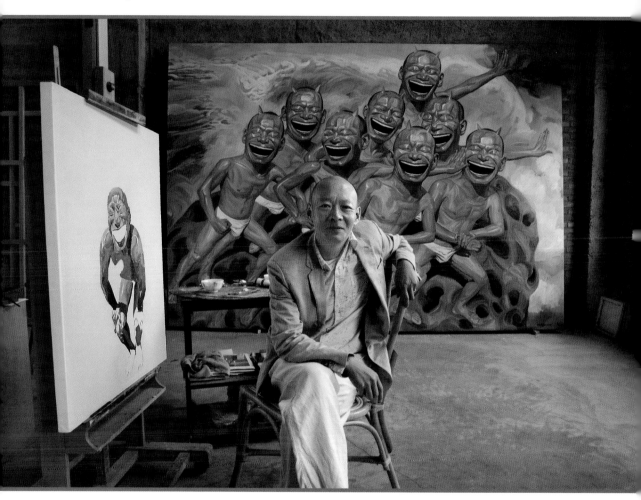

I first met Yue Minjun in 2000 in a relatively small studio he used in the Yuanmingyuan artists' village. His paintings of the laughing Chinese man were already becoming quite well known and I saw for the first time, the life-size sculptures he was creating of the same character. I remember how much he resembled his subjects, yet how serious he was in conversation. The photographs I made of him at that time just didn't seem to blend the man and the work. Recently, I visited Minjun in his new home in a suburb of Beijing. As we entered his studio, I saw immediately where I wanted to photograph him, in front of a large canvas with multiple figures. I positioned him in a chair in front of an easel with an unfinished painting so that he would blend into the canvas in the background. Everything seemed perfect except Minjun was again fairly solemn. As we began to talk about how very far he has come in his career and how well known he is eight years later, he slowly became more like his laughing man and I knew within the first two photographs I made, that I had the portrait I wanted.

The studio of Yue Minjun

告知
园区内大面积施
实行内部交通管

Entrance to 798 arts district, 2007

A street in the 798 arts district, 2007

Wang Guangyi sculpture in 798 arts district, 2007

798 Space gallery, 2004

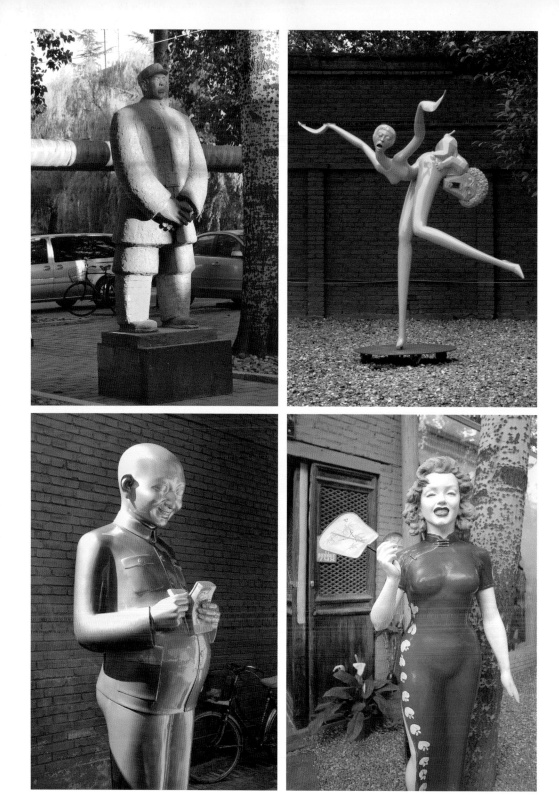

[Upper Left] Liu Ruowang sculpture at 798 arts district, 2008

[Upper Right] Ren Zhihui sculpture at 798 arts district, 2008

[Lower Left] Ren Hongwei sculpture at 798 arts district, 2008

[Lower Right] Sui Jianguo sculpture at the 798 arts district, 2008

Sculptures by Sui Jianguo (foreground) and Yue Minjun (background) at 798 arts district, 2008

[Left] Installation at the Ullens Center for Contemporary Arts in the 798 arts district, 2008

[Right] Sculpture in a gallery at the Songzhuang Artist Village, 2008

[Left] Performance artist at 798 arts district, 2007

[Right] Venus de Milo as a logo sculpture in the Liquor Factory arts district, 2007

Walking
Al...
走在线上

The studio of Wang Guangyi, 2000

[Left] Sculpture at the Concave & Convex gallery space in the Songzhuang Artist Village, 2008

[Right] "Demolition" by Zhang Dali in the Songzhuang Artist Village, 2008

Acknowledgements

I dedicate this book to my wife Cyndy and my daughters Aija and Kari who have always graciously encouraged me to return to China to pursue my work. Only they know what their support has truly meant to me.

I owe a debt of gratitude to Robert A. Hefner III who first introduced me to China thirteen years ago. His knowledge of history and world affairs and his belief in China's future had a profound influence on me.

I want to thank my friend and associate Yang Jie who has been with me on practically all of my days in Beijing. Without her knowledge of the city and her ability to negotiate its many levels, I would not have discovered all that I have about it. Her guidance and understanding of my work was of great importance. I would also like to acknowledge Jie's husband Weiqiang Han who became an equally enthusiastic advisor. I always enjoyed our photographic explorations together and I look forward to many more in the future.

I want to acknowledge Janet Brewer for her thoughtful and thorough editing of my text. Most importantly, I want to thank the editorial staff of Foreign Languages Press including Liu Fangnian, Yang Chunyan, Hu Kaimin, and Nicole Ouyang. Their interest in my photographs and their hard work devoted to creating this book is greatly appreciated. The production of *Beijing Days* is a combined achievement that I am most pleased to share with them!

Jon Burris

图书在版编目（CIP）数据

在北京＝Beijing Days：英文／（美）布里斯
（Burris, J.）著． —北京：外文出版社，2008
ISBN 978-7-119-05488-9

Ⅰ．在… Ⅱ．布… Ⅲ．北京市—概况—英文 Ⅳ．K291

中国版本图书馆CIP数据核字（2008）第136985号

英文审定：梁良兴
责任编辑：刘芳念
装帧设计：视觉共振设计工作室
印刷监制：张国祥

在北京

Jon Burris 著

©2008外文出版社

出版发行：

外文出版社（中国北京百万庄大街24号）

邮政编码：100037

网址：http://www.flp.com.cn

电话：008610-68320579（总编室）
　　　008610-68995852（发行部）
　　　008610-68327750（版权部）

印刷：

北京外文印刷厂

开本：787mm×1092mm　1/16　印张：10

2008 年第 1 版　第 1 次印刷

（英）

ISBN 978-7-119-05488-9

10500（平）

85-E-669P